101

Cat Care

101
ESSENTIAL TIPS

Cat Care

Penguin
Random
House

Produced for DK by Sands Publishing Solutions
4 Jenner Way, Eccles, Aylesford, Kent ME20 7SQ

Editorial Partners	David & Sylvia Tombesi-Walton
Design Partner	Simon Murrell

Senior Editor	Chauney Dunford
Senior Art Editor	Elaine Hewson
Managing Editor	Penny Warren
Jacket Designer	Kathryn Wilding
Production Editor	David Almond
Production Controller	Rebecca Parton
Art Director	Jane Bull
Publisher	Mary Ling
Special Sales Creative Project Manager	Alison Donovan

Written by	Sylvia Tombesi-Walton
Consultant	Kim Bryan

First published in Great Britain in 2015 by Dorling Kindersley Limited
DK, One Embassy Gardens, 8 Viaduct Gardens, London SW11 7BW

Copyright © 2015 Dorling Kindersley Limited
A Penguin Random House Company
4 6 8 10 9 7 5
011–274501–May/2015

A CIP catalogue record for this book is available from the British Library.
ISBN 978-0-2410-1471-4

Printed and bound in China

For the curious
www.dk.com

101 ESSENTIAL TIPS

DECIDING ON A CAT

1 IS A CAT RIGHT FOR YOU?

Mysterious and entertaining, cats have a reputation for being independent, but they still rely on us for a range of needs. These include food, shelter, and veterinary care, but also affection and physical and mental stimulation. Cats will adapt easily to living indoors, but it is important to keep them entertained or they may get bored and develop bad habits such as scratching furniture or chewing plants.

PLAYING TOGETHER
Cats learn how to interact with other cats and with people during the first few weeks of their lives. This early socialization period is extremely important.

2 ARE YOU RIGHT FOR A CAT?

Before welcoming a cat into your home, be aware that costs – food, veterinary expenses, pet insurance, boarding when you go away – will add up. Consider your lifestyle, too: do you have the time and space to offer a suitably stimulating environment that caters to all of your cat's needs?

A LIFETIME OF LOVE
That kitten in your hands might live 15 years or more. You are responsible for its wellbeing.

3 PEDIGREE OR CROSS-BREED?

If you wish to enter the world of cat shows, buy a pedigree cat from a reputable breeder. Pedigree cats are expensive, but by opting for a specific breed, you can expect certain physical and behavioural characteristics. If all you want is a low-maintenance companion, then a non-pedigree shorthaired cat is probably your best option.

Longhaired cats need regular grooming sessions

Silver Tabby Maine Coon

Cross-breed with tabby markings

4 KITTEN OR ADULT CAT?

It can be easier to introduce a kitten rather than an adult cat into your home, especially if you already have a pet. However, adult cats tend to be less demanding and do not usually require any training. In addition, an adult cat is likely to have already been neutered, reducing the initial costs.

Kittens have more fun when they have playmates

Adult cat comes with fully developed personality

Inquisitive kittens

Young adult cat

5 ONE CAT OR TWO?

Two cats cost more than one, of course, but if you go out to work during the day, two cats can keep each other company, if you have room. Be sure to neuter cats of the opposite sex that live together, to avoid unwanted kittens. Also, an adult cat that is used to living alone might not welcome a feline companion.

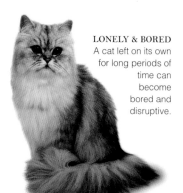

LONELY & BORED
A cat left on its own for long periods of time can become bored and disruptive.

FRIENDS FOREVER
Bonded kittens will remain best friends for life. They will enjoy each other's company during playtime and often cuddle up together to sleep.

6 MALE OR FEMALE?

In general, female cats are more affectionate and playful than male cats, who are more likely to show territorial aggression. However, it should be personality, rather than gender, that dictates your choice of a cat. This is especially true after spaying or neutering, when any character differences between males and females tend to become less evident.

Cats of opposite sex often get along best

Bonded male and female cats

Adult male cats are often solitary creatures

7 LEGAL RESPONSIBILITIES

In the United Kingdom, cats are considered as having a right to roam. The law accepts that a cat cannot be restrained or managed at all times when outside of your home. As a consequence, cat owners do not have legal responsibility for the actions of their cats. However, they are expected to take the necessary measures to ensure that their cat does not cause injury to people or damage to property.

YOUR RESPONSIBILITY TOWARDS YOUR CAT

Under the Animal Welfare Act 2006, cat owners have a duty of care towards their pet and are responsible for its welfare. This means they have to provide, among other things, a proper diet, a suitable environment, the opportunity to exhibit normal behaviour, and protection from pain, suffering, and injury.

8 INDOOR OR OUTDOOR CAT?

Cats do enjoy exploring the great outdoors, but they face many dangers there. These include speeding vehicles, parasites and infectious diseases, predators, and even other cats. Indoor cats are healthier and live longer than their outdoor friends. However, it is important to keep them mentally and physically active by engaging them in regular play sessions.

ARMCHAIR HUNTER
Make sure you give your indoor cat access to a window. It will enjoy spending time looking at the outside world from the safety of its home environment.

KING OF THE GARDEN
Before you let your cat roam outdoors, make sure it is safe for it to do so. If your garden is already part of another cat's territory, you may need to accompany your cat out at first.

9

WHICH BREED? PLAYFUL CATS

Through playing, kittens learn the skills they need to interact socially and to hunt. Some cats carry their love of playtime into adulthood. Particularly energetic cat breeds include Abyssinians, Devon or Cornish Rexes, and Siamese. Neutering your cat before it reaches sexual maturity ensures that it will remain playful into its adult life.

Preparing to "attack" litter mate

TINY ACROBATS
Watching two kittens at play is endlessly entertaining. They pounce, jump, stalk, and leap in an attempt to learn the skills they will need later in life.

Watching and preparing for next move

10

WHICH BREED? DOCILE CATS

If the idea of having regular playful sessions with your cat does not appeal, you could opt for a breed with a low level of playfulness. Some breeds – including Ragdolls, Longhairs (or Persians), and British Shorthairs – are especially docile and laid-back. The Ragdoll, in particular, is famous for its tendency to relax completely and go limp in the arms of a trusted person. If you wish to keep your cat indoors, it might be best to choose one of these breeds.

CHILLING OUT
Ragdoll cats such as this one enjoy a relaxing environment.

11 PHYSICAL VARIATIONS

There are many cat breeds, each with a different set of physical characteristics. While some of these attributes have evolved naturally, others have come about as a result of selective breeding. Variations include coat length and colour, body shape, and eye shape and colour.

FACE & EYES
Cats' faces can be wedge-shaped or round; their eyes can range in colour from orange and amber, to green and blue.

BODY SHAPE
Cats' body shapes have evolved to suit the climate in their places of origin. Stocky breeds usually hail from cold countries, while lithe breeds often originate from countries with a warm climate.

Muscular

Slim

Stocky

12 COATS: LONG, SHORT & SPECIAL

Cats are loosely divided into two groups: longhaired and shorthaired. However, some breeds do not fall into either camp, such as the hairless Sphynx and the various Rex breeds, with their wavy, rippled hair. All cats produce proteins in their saliva, skin, and urine that are transferred to their coats and can cause reactions in some people; if you or someone in your home is so afflicted, consider one of the so-called hypo-allergenic breeds.

LONG COAT
A longhaired cat's coat needs daily brushing and regular grooming, or it will become matted.

SHORT COAT
Shorthaired cats tend to groom themselves, but you can help them with regular brushing.

HAIRLESS
Although known as a hairless cat, the Sphynx is actually covered with a layer of fine down.

REX COAT
Devon, Cornish, and Selkirk Rex cats have a wavy coat that is silky and soft to the touch.

13

13 BUYING FROM BREEDERS

If you have done your research and know which breed you want, the next thing to do is contact a specialist breeder. This is particularly important if you wish to show your cat or to breed from it. A reliable breeder will be able to identify potential winners within a litter. When you buy a pedigree kitten from a breeder, ask if you may see its mother, too. Good breeders will always be happy to oblige.

SIBERIAN FOREST
These Siberian Forest kittens are still too young to leave their mother. A good breeder will wait until the kittens are at least 12 weeks old before releasing them.

Cat pen at a rescue centre

14 RESCUE CENTRES

Cat shelters are full to bursting with abandoned cats and kittens in need of loving homes. If you are looking for a feline companion rather than a show cat, consider adopting from a rescue centre. Any adult cats will have been neutered already, reducing your initial expense. You may be asked for a donation towards the running of the shelter.

15 FRIENDS & ACQUAINTANCES

Another way of obtaining a cat is by spreading the word among your friends and family. Somebody might know someone else whose cat has just had a litter and who is hoping to have the kittens adopted into a good home. Your veterinary surgery is another good source of information.

Litter mates cosying up together

16 ADOPTING A STRAY

If a stray-looking cat keeps visiting your garden and you wish to give it a home, start by leaving some food out for it. If it is approachable, take the cat to the vet to establish whether it is indeed a stray. The vet will be able to check if the cat is microchipped and simply lost. If the cat is a stray, have it examined for infectious diseases and vaccinated, and find out if it has been neutered, before taking it into your home. This is especially important if you already have cats.

GENTLY, GENTLY
Slowly approach cats that seem to be strays – they may be fearful of people and react by scratching or hissing. Better yet, sit and let them come to you in their own time.

STRAY OR FERAL?
While strays are often abandoned or lost cats that have lived with a human family for some time, feral cats have had no significant interaction with humans. They are often nervous and unapproachable.

GAINING TRUST
Leave some food out for the stray cat. At the start, it might not eat until after you have left, but it will soon come to think of you as a provider of food. Move the bowl a little closer to your door each day.

YOUR NEW FRIEND

17 CAT-PROOF YOUR HOME

Before you bring your new cat to live with you, make sure that your home provides a safe environment for it. Cats are curious by nature, so invest in a few child-proof latches to prevent them from exploring food cupboards, china cabinets, and rubbish bins. It is also wise to purchase some safety covers for any electric sockets.

Make sure windows are closed or on a latch to prevent accidental falls

Keep door of washing machine closed at all times

Put cat's litter tray in an out-of-the-way corner

Make sure cat's food station is in a quiet area

18 DANGERS IN THE HOME

Some substances routinely found in the home are toxic for cats and could be potentially fatal if ingested. These range from household cleaning products and antifreeze, to human painkillers and some decorative house plants and flowers, such as poinsettia, mistletoe, and lilies. Long cords from curtains and blinds may also be hazardous for inquisitive cats, as can needles, thread, and rubber bands.

19 CAT-PROOF YOUR GARDEN

Wait until your cat is familiar with and comfortable in its new home before allowing it to go outside. This could take a couple of weeks. Use this time to cat-proof your garden. Make sure there are no holes or gaps in the fences; while most cats will be able to climb over, a fence ought to prevent dogs, foxes, and other predators from coming into your garden.

20 DANGERS IN THE GARDEN

The dangers in the garden are similar to those in the home – that is, toxic substances such as weedkillers, pesticides, and slug and snail pellets. Close every container carefully, and store it in a cat-proof shed. Also use the shed to store sharp and heavy gardening tools. Other hazards include toxic plants such as ivy, rhododendron, and clematis.

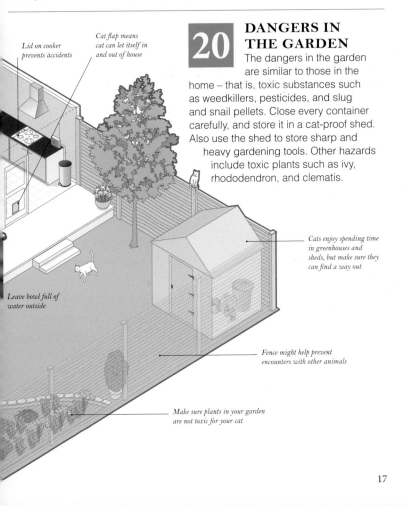

Lid on cooker prevents accidents

Cat flap means cat can let itself in and out of house

Cats enjoy spending time in greenhouses and sheds, but make sure they can find a way out

Leave bowl full of water outside

Fence might help prevent encounters with other animals

Make sure plants in your garden are not toxic for your cat

17

21 BASKETS

Cats spend most of the day sleeping or napping, so make sure their sleeping quarters are cosy, comfortable, and in a quiet area of the house. Cats enjoy feeling snug and warm, so consider buying a basket with high, padded sides and a cushioned centre that, ideally, is removable and washable.

Igloo basket

BASKET CHOICES
There are many types of cat basket on the market – from igloos, to wicker and padded varieties.

Cat basket

Cushioned basket

22 BEDDING

Choose bedding material that is comfortable, soft, and machine-washable. Cats love being warm, so for the winter months, consider buying a fleecy bed that hangs from the radiator, allowing your cat to be as close to a heat source as possible. If your radiators are not suitable for this type of bed, a self-warming bed is the next best thing. Many cats enjoy nothing better than sleeping on a blanket on your lap.

Kitten snoozing on fleece bedding

23 FEEDING BOWLS

Cats are fussy eaters with an extremely well-developed sense of smell. Keep their eating station clean and wash their feeding bowls at least daily, especially in the warm summer months. Feeding bowls come in a range of materials – from plastic and metal, to ceramic. If buying a plastic or metal bowl, make sure it has rubber feet to prevent it from sliding around the floor.

Plastic bowl

Metal bowl

24 LITTER TRAY

There are two types of litter tray: open and covered. Choose your favourite and stick with it, because a cat that is used to an open tray may feel nervous about switching to a covered one, and vice versa. Place the tray in a quiet corner, well away from your cat's feeding station. Scoop out any stools and clumps of urine several times a day to keep the tray as clean as possible.

Poop scoop

25 TYPES OF LITTER

There are many types of litter available. As with the litter tray, try to stick with whatever your cat has grown used to. Clumping litter is convenient in terms of scooping out faeces and urine; however, what is appealing to us is not always acceptable to our cats – for example, perfumed litter often deters cats from using the litter tray altogether.

Non-absorbent litter

Fuller's earth litter

Lightweight litter

Wood-based litter

26 CAT FLAP

If your cat is an indoor/ outdoor cat, have a cat flap installed. They range from simple models that open inwards and outwards, to others that feature a locking dial. This allows you to leave the flap fully open, fully closed, or so that it opens only inwards; this way, the cat can enter but not leave the house. The most sophisticated allow a cat through based on reading its microchip.

Magnetic strip

Locking dial

IN & OUT AS THEY PLEASE
A cat flap allows a cat the freedom to come and go, but the locking dial gives cat owners ultimate control.

27 GROOMING EQUIPMENT

Most cats enjoy being brushed. However, a regular brushing routine is vital for longhaired cats, because it helps them maintain a glossy, tangle-free coat. The most important tools are a toothed comb, a slicker brush with soft wire bristles, and a soft brush. Accustom your cat to having its teeth brushed from an early age.

TOOLS OF THE TRADE
You will need combs and brushes, a toothbrush, and cotton-wool balls for cleaning ears and eyes.

EMERGENCY ALTERNATIVE
Cat claw clippers are the best and most accurate tool to use, but human clippers may be used with care if necessary.

28 CLAW CLIPPERS

Get your cat into the habit of having its claws trimmed regularly. This is especially important for indoor cats, which are less likely to wear down their claws. Press gently on each paw pad to expose the claw, then, using cat claw clippers, cut just the white tip of the nail. Do not cut to the quick; this will cause bleeding and may risk infection if not treated quickly.

Cat claw clippers

29 CARRY CASES

Most cats do not take naturally to walking on a lead, so a cat carrier is essential. Carry cases come in various materials: wicker, plastic, cardboard, or wire. Because you will mostly use a carrier for trips to the vet, your cat might associate it with bad experiences. Prevent this from happening by turning it into a safe place for your cat: leave it out at all times, maybe next to a radiator, and put a warm blanket in there.

Cat carrier

SCRATCHING POSTS

30 Cats scratch for a variety of reasons, not least marking their territory and keeping their claws sharp. To help prevent your cat from scratching destructively around the home, provide it with a scratching post. Some scratching posts double as multifunctional units, with sleeping areas, hanging toys, and other features.

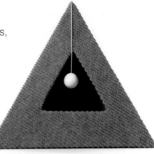

Pyramidal scratching post

Scratching post with platform

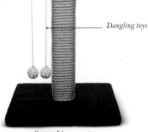

Dangling toys

Scratching post

TOYS

31 It is important to keep your cat physically active and mentally stimulated. Provide a selection of toys to chase, swat, pounce on, and bite to re-create the feel and thrill of the hunt and keep your cat active, youthful, fit, and happy. Interacting with your cat at playtime will also strengthen the bond between you.

Plastic ball

Rolling toy

Fluffy toy

Catnip mouse

21

32 CHECKING THE HEALTH OF A NEW KITTEN

A healthy cat could be your companion for 15 years or more. If you are getting a kitten, choose carefully and spend some time with it observing its behaviour before bringing it home. Consider the way it interacts with you and its siblings. Vitality and playfulness are important indicators of good health, as is an inquisitive attitude. There are also a few key checks to help establish the overall wellbeing of a kitten.

Ears
If the kitten is scratching its ears obsessively, it may have ear mites. The presence of mites is also revealed by a build-up of dark-coloured wax. A healthy kitten should have clean, dry ears.

Nose
Make sure there is no discharge in or around the nostrils. This could be a sign of ill health. The nose should feel cool and moist.

Eyes
Eyes should be bright, clear, and free of discharge. If the kitten's third eyelid is showing, there might be an underlying health issue.

Mouth
Healthy gums are pale pink in colour. The roof of the mouth should also be pale pink. A kitten's breath should not be unpleasant.

Coat
A kitten's coat should be glossy, thick, and soft. Check for fleas by parting the fur to expose the kitten's skin, and look for black specks.

Legs and paws
Legs should be strong and straight. Watch the kitten as it walks and plays, and make sure there is no lameness.

Abdomen
The belly should be toned and rounded but not swollen. A pot belly usually indicates the presence of worms.

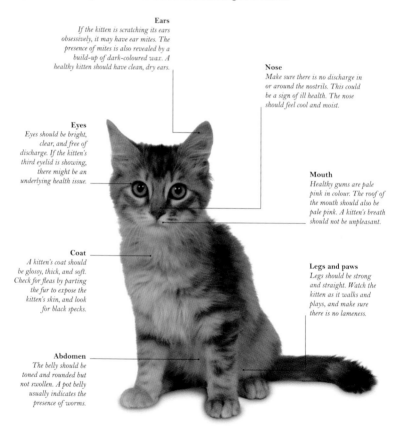

33

CHECKING THE HEALTH OF AN ADULT CAT

It is a good idea to learn to recognize the indicators of feline ill health and to acclimatize your cat to regular home examinations. This will allow you to catch any potential problems at an early stage. Carry out a home examination when your cat is relaxed, perhaps while he is on your lap, and over the course of several days. Reward your cat's cooperation with a treat or a cuddle. If anything appears wrong, take your cat to the vet. Outdoor cats should be checked more often than indoor ones.

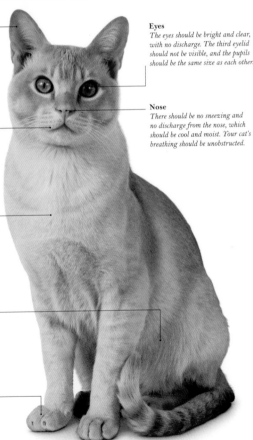

Ears
An adult cat's ears should be clean, dry, and free of unpleasant odours. If your cat holds its head to the side or shakes it and scratches its ears, there may be an infection or mites present.

Eyes
The eyes should be bright and clear, with no discharge. The third eyelid should not be visible, and the pupils should be the same size as each other.

Nose
There should be no sneezing and no discharge from the nose, which should be cool and moist. Your cat's breathing should be unobstructed.

Mouth
Your cat's gums should be pale pink, its teeth white and strong, and its breath not unpleasant. A build-up of tartar and inflamed gums might mean you have to improve its oral hygiene.

Coat
A cat's coat is a great indicator of good health. It should be glossy and thick; if it sits up in patches or becomes dull, take your cat to the vet. Also look for bald patches, scratches, excessive shedding, and parasites.

Abdomen
Run your hands down the side of your cat. You should be able to feel an indent at its waist. If your cat's stomach is round and protruding, it might be overweight or pregnant.

Paws
Your cat should distribute its weight through all four legs and show no signs of discomfort when it walks or jumps.

23

34

SETTLING IN A NEW KITTEN

Bring your kitten home at a time when you can spend a couple of days bonding with it. Bear in mind that it will be away from its mum and litter mates for the first time in its life, so create an environment in which it will feel calm and secure. For the first few days, do little beside sitting in the room with it, ideally at floor level, so the kitten can investigate and get to know you.

1 Place the carrier in a quiet room and open the door. The kitten will probably observe its surroundings from the safety of the carrier.

Children can help, but they ought to be supervised

CHILDREN & KITTENS

Your children will be excited about the new arrival. Allow them to help settle the kitten in, but make sure they maintain a calm, quiet demeanour.

2 Give the kitten time to build up its confidence. It will come out of the carrier in its own time and start exploring the room.

Kitten is attracted to food

35

SETTLING IN AN ADULT CAT

Coming from a rescue centre or another family, an adult cat might find settling into your home as stressful as a kitten would. The same approach applies: do not force the cat out of the carrier, but let it come out in its own time and allow it to explore just one room at the start. Create a calm environment, with a couple of places for the cat to hide should it feel threatened or nervous. Get to know each other.

INITIATING CONTACT
Let the cat explore its surroundings without interference. Allow it to initiate contact with you, which it will do when it feels confident.

Big pupils in bright light indicate fear

FEARFUL BEGINNINGS
In new, unfamiliar surroundings, a cat might display fearful behaviour, such as hiding under the bed, crouching low to the floor, or walking close to the walls. Do not add to its fear by trying to approach it. Give it time.

Walking low to the floor

36 CHOOSING & TEACHING A NAME

If you bought your cat from a breeder, it will have a registered pedigree name, which is often quite a mouthful. In everyday life, your cat's name should be one or two syllables at most, such as Homie, Rudy, Lyra, or Zack. To teach your cat its name, repeat it as you give treats. A cat adopted from a rescue centre may already have a name, and staff there may advise you to keep it.

POPULAR NAMES

As TS Eliot suggested in his poem on the subject, it is no easy task to name your feline friend. You might want to get to know your cat a little before giving it the name that will accompany it for the rest of its life.

A cat's name may be inspired by a particularly strong personality trait or by a physical characteristic – a white cat may be called Snowy, for example; a black one, Sooty; and a cat with only half a tail might be Stumpy.

Some people choose to name their cats after their favourite actor or pop star.

SHORT ATTENTION SPAN
Kittens are hard to train because they are easily distracted.

37 TEACHING A CALL SIGN

Training sessions with cats should last just a few minutes and ideally take place before mealtimes. Say, "Come!" and shake a pack of your cat's favourite biscuits. When it comes, give it a treat. Move away and repeat this exercise a couple of times. Cats respond to positive reinforcement and will soon learn to come to the call sign even without the rattling sound of the snack pack.

REWARD TRAINING
Cats are very intelligent and trainable creatures. The secret is to find a reward they enjoy and that keeps them motivated to learn, whether it be a food treat or a cuddle.

38 MICROCHIPPING

The most reliable way of identifying a lost cat and having it reunited with its family is by microchipping it. Unlike collars and tags, which can come undone, the microchip is permanently embedded under the cat's skin. A microchip is about the size of a grain of rice. Each has a unique code that is linked to the contact details of the cat's owner in a database.

INSERTING THE MICROCHIP
The microchip is inserted in the back of a cat's neck by means of an injection. The cat will have no awareness of it.

SCANNING FOR A MICROCHIP
When a stray cat is taken to a vet or a rescue centre, it is routinely scanned for a microchip to see if it has a family.

39 VACCINES

Have your kitten vaccinated against three potentially fatal viruses: feline infectious enteritis, feline influenza, and feline leukaemia. You are advised to do this even if your cat is indoor only. You may have to use a cattery one day, and vaccinations are mandatory for residents. If you are acquiring an adult cat from a shelter, confirm that its vaccinations are up to date.

KITTEN'S FIRST VACCINATION
Kittens are usually vaccinated at between eight and ten weeks. Throughout its life, a cat will also require annual boosters.

40 NEUTERING

Unless you have a pedigree cat that you wish to breed, it is wise to neuter your cat. Male cats should be castrated at about four months to prevent undesirable behaviour such as roaming, fighting, and spraying. Female cats can be spayed as early as four months to avoid unwanted pregnancies. Neutering is a routine operation that is carried out under anaesthetic.

Relaxed cat

41 PICKING UP A CAT

You may need to gain the trust of an adult cat before you can pick it up. Let it come to you and sniff you; then pet it gently. Put one hand under the cat's chest, and support its hindlegs with the other. Lift it up slowly, all the time holding it close to your chest, so it feels secure.

KNOW WHEN TO PICK UP

Be mindful of cat body language. Once you have your hand under its chest, only proceed to pick up the cat if it remains calm.

42 HANDLING A CAT

After you have picked up your cat, cradle it gently against your chest. Create as many points of contact as you can between your body and your cat's. This will increase its feeling of safety. If a cat starts struggling while in your arms, put it back down as quickly and gently as you can. Try to maintain a hold of the cat without getting scratched until it is safely on the ground.

FEELING SNUG

With firm support around and under its hindquarters, this cat looks relaxed in the arms of its owner. Stroking the cat while holding it also reassures it.

Inquisitive, alert look

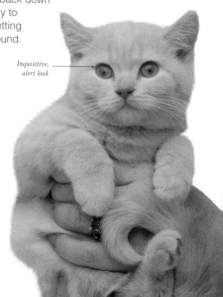

43 MEETING OTHER CATS & PETS

Keep the new cat and the established pet in separate rooms at first, switching them around every few days, so they can become familiar with each other's smell. Supervise the first few meetings, making sure the new arrival has a safe place to retreat to in the event of an aggressive reaction.

Curious dog

CAT MEETS DOG
Contrary to popular belief, cats and dogs can become good friends, especially when introduced to each other at a young age.

Apprehensive kitten

First encounter between an adult cat and a kitten

---

44 MEETING CHILDREN & FRIENDS

Children can often be too loud and boisterous for cats and kittens, so supervise early encounters until you are confident that both are comfortable in each other's presence. As ever, allow the cat to make the first move, and do not force it to be in the presence of any house guests.

SITTING PRETTY
The safest approach with young children is to have them sit on the floor with a cushion on their lap. Gently place the kitten on the cushion.

CAREFUL HANDLING
Make sure children pick up and handle the cat in the correct manner.

CARING FOR YOUR CAT

45 WET & DRY FOOD

Commercial cat food is scientifically formulated to provide your cat with a balanced diet that meets all of its nutritional needs. Kittens and senior cats, which have slightly different dietary requirements from adult cats, have their own ranges of food. To add variety to your cat's diet, buy wet food in a range of flavours and supply it in conjunction with dry biscuits and fresh food. Introduce any dietary changes slowly to avoid stomach upsets.

DRY FOOD
These mini-biscuits come in various flavours and are good for keeping your cat's teeth free of tartar.

Boneless fish flakes

Chicken and turkey

Tuna chunks

Lamb chunks

BISCUIT BOWL
This cat appears to have a preference for dry food. Some biscuits are particularly suited to specific dietary needs – for example, dealing with hairball relief or weight control.

WET FOOD
A stunningly diverse range of flavours – from fish, to poultry, to meat – ensures that no cat ever has to grow bored with the food in its bowl.

46 PROVIDE ENOUGH WATER

Cats are desert animals with a relatively low thirst drive, and they draw most of the water they need from the food they consume. However, it is important to provide your cat with a bowl of fresh water at all times, especially if it has a predilection for dry food, which has a very low moisture content. Some cats prefer to drink running water, and they might make a beeline for a tap when it is turned on.

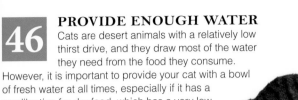

Drinking water from a bowl

47 CAT TREATS

There are many types of cat treat on the market, so choose wisely. Avoid treats with a high fat content, and instead opt for those that are high in protein, such as dried fish flakes. Treats can also help keep teeth and gums healthy. The key word with treats is moderation.

TREAT RAINBOW
Like wet and dry food, cat treats come in a wide range of flavours, as well as different shapes and textures.

48 GIVING MILK

One of the most persistent feline myths is that cats love a saucer of milk. However, not only is milk not essential to a cat's diet, but many cats are actually lactose intolerant, so drinking cow's milk can lead to a stomach upset. The only exception concerns young kittens, who need their mother's milk to grow up and thrive. If the mother is not around or is not producing enough milk, you will have to feed the kitten a special kitten formula.

Drinking specialist kitten milk

49 HOW MUCH FOOD & WHEN?

Cats are obligate carnivores, which means they need meat to survive. Unlike dogs, cats are not usually greedy, and they will adjust their calorie intake to their level of activity. As a result, as a cat grows older and less active, it might start to eat less. If you are concerned, ask your vet for advice on an older cat's food intake.

MULTIPLE BOWLS
In order to prevent competitive eating, it is advisable to give each cat in your household its own feeding bowl.

FEEDING GUIDE
This chart is an estimate of daily feeding requirements (in calories, as well as number of grams of both wet and dry food) of your cat, based on its current weight and lifestyle.

Adult weight	2kg	4kg	6kg	10kg	12kg
Inactive cat	100–140kcal (120g wet/ 30g dry)	200–280kcal (240g wet/ 60g dry)	300–420kcal (360g wet/ 90g dry)	400–560kcal (480g wet/ 120g dry)	500–700kcal (600g wet/ 150g dry)
Active cat	140–180kcal (160g wet/ 40g dry)	280–360kcal (320g wet/ 80g dry)	420–540kcal (480g wet/ 120g dry)	560–720kcal (640g wet/ 160g dry)	700–900kcal (800g wet/ 200g dry)
Pregnant female	200–280kcal (240g wet/ 60g dry)	400–560kcal (480g wet/ 120g dry)	600–840kcal (720g wet/ 180g dry)	800–1,120kcal (960g wet/ 240g dry)	1,000–1,400kcal (1,200g wet/ 300g dry)

50 SETTING A ROUTINE

Some people work long hours outside the home and resort to free-feeding their cats. This means the cat has food at its disposal and can eat when it is hungry. In such cases, use dry food, which doesn't go bad in the heat. However, if your lifestyle allows it, opt to establish a feeding routine for your cat. Feed it two meals a day, at regular times, but leave a bowl of water out at all times.

Enjoying a meal at dinner time

51 WHERE TO FEED

Create a feeding corner in an out-of-the-way area of the kitchen without much foot passage. Cats have a very keen sense of smell, so keep their eating area clean. If you have a multiple-cat household, give each cat its own bowl. This will prevent squabbles over food and make it easier for you to track each cat's eating habits.

Obese and unhappy

52 OVERFEEDING

Be careful not to overfeed your cat, especially if it is an indoor cat without many opportunities to burn calories. Overfeeding could lead to obesity, which brings with it a host of health issues, ranging from difficulty grooming and joint problems, to potentially life-threatening conditions such as diabetes and heart disorders. If you are not overfeeding your cat but it is still putting on weight, it could be due to a hormonal imbalance. Consult your vet.

53 DEALING WITH A FUSSY EATER

If your cat becomes a fussy eater, try feeding it a different flavour or a different brand of wet food. Keep a list of the flavours it likes. Consider the cleanliness of the feeding station. Remove any scraps of old food, and wash the bowl before refilling it. If your cat still will not eat, there may be an underlying health issue. Talk to your vet.

SNIFFING FROM AFAR
Feed your cat strong-smelling food to attract it to its bowl.

54 WHY CATS EAT GRASS

Although grass has no real nutritional value, it plays an important role in your cat's diet. Cats eat grass because it facilitates the bringing up of indigestible matter sitting in your cat's stomach, such as hairballs. If your cat has no access to a lawn, you should provide it with some indoor greenery. Beware, though: some plants may be toxic to your cat. Stick with special kitty grass or opt for cat-friendly herbs like thyme, sage, or parsley.

Munching on grass

CREATE A KITTY LAWN
You may not have a lawn, but it is easy to grow some cat-friendly grass in a pot. Your cat will nibble at it as it feels the need – and thank you by bringing up a furball!

55 GROWING CAT GRASS

Grow-your-own-grass kits can be found in most pet shops and well-stocked supermarkets. Just follow the simple directions on the packet, and within a week you should have a miniature grassy lawn for your cat. The grass in these containers keeps well for up to three weeks. Alternatively, devote a patch of your garden to plants that your cat might enjoy (see Tip 54).

Thick, luscious grass

Foil container

Grow-your-own-grass kit

56 GROOMING LONGHAIRED CATS

Longhaired cats need a helping hand with grooming. Daily brushing sessions prevent the formation of painful knots and mats in the fur and help remove loose hairs. Without brushing, these hairs would be ingested by your cat, leading to large furballs in its stomach and subsequent regurgitation. A good bonding exercise, grooming is also beneficial to the cat's circulation and helps distribute natural oils all over its coat, which helps keep it in good condition. Another advantage is that the regular handling of your cat allows you to become aware of any abnormalities or lumps on its body.

1 Brush your cat when it is feeling calm and relaxed. Start by running a slicker brush all down its back and sides. Be gentle, and do not pull – there might be knots in the fur that need to be untangled.

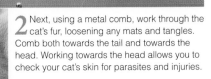

2 Next, using a metal comb, work through the cat's fur, loosening any mats and tangles. Comb both towards the tail and towards the head. Working towards the head allows you to check your cat's skin for parasites and injuries.

3 The belly and the legs are very sensitive. If there are knots, loosen them with your fingers rather than the metal comb. Finish on the tail with a wide-toothed comb or brush.

57 GROOMING SHORTHAIRED CATS

Although shorthaired cats do not require the same level of brushing as longhaired cats, it is advisable to establish a weekly grooming routine to keep their coats glossy and in good condition. Regular handling also allows you to spot any abnormalities that might otherwise go unnoticed. Operate in a calm, relaxed environment, and watch out for any signs of nervousness or discomfort from your cat during the session.

CHILD'S PLAY
Brushing a shorthaired cat is considerably easier than brushing a longhair. At its simplest, it is just a matter of running a brush gently down its coat to remove loose hairs.

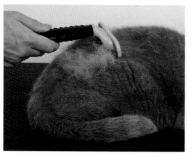

1 Start by working a slicker brush down the back and sides of your cat. If it allows you, brush its abdomen, too. Be gentle, though, since this is a very sensitive area.

2 Using a metal comb, part the fur to check for fleas, scratches, and other injuries. If you like, apply a few drops of coat conditioner to enhance the shine of your cat's coat.

REGULAR STROKING
Some cats do not enjoy being brushed and combed. However, even regular stroking goes some way towards ensuring your pet maintains a healthy, glossy coat.

In addition, there are benefits for us, too, since stroking a cat has been shown to lower blood pressure and relieve stress.

58

CLEANING CATS' EYES

Some cats, especially longhaired cats with shortened faces – Longhairs (or Persians) and Himalayans, for example – need a little help to keep their eyes clean. This is because their tear ducts are particularly prone to blocking up, leading to a build-up of gunk. If left for even just a day, this discharge can turn crusty, which makes it still harder to remove. Eventually, it can even lead to an unsightly discoloration of the fur around the eyes and nose. If you notice an unusual crusty build-up around your cat's eyes, there may be an underlying health issue, so speak to your vet.

REGULAR INSPECTION

A regular examination of your cat's eyes helps catch problems at an early stage. A visible third eyelid is a sign of ill health or injury; the latter is more likely if only one eye is affected. If your cat keeps an eye closed, it might have been scratched or have a foreign body in it.

1 Dampen a piece of cotton wool with lukewarm water or, if the build-up seems particularly stubborn, with a saline solution such as you would use for contact lenses. Wipe from the nose towards the temple.

2 Repeat the process on the other eye with a clean piece of cotton wool. If your cat allows it, try to remove any staining on its fur, too. Avoid your cat's eyeball during cleaning, since it could lead to irritation.

59 CLEANING CATS' EARS

Although cats are perfectly capable of looking after their ears, a regular cleaning routine has the added benefit of allowing you to keep an eye out for parasites, scratches, and other issues (see right). Outdoor cats, for example, might come home with barbed seeds or spiky grass stuck inside their ears. Foreign bodies should be removed from your cat's ears only if they are on the surface. Anything lodged in the ear canal should be removed by the vet. To clean your cat's ears, you will need some cotton wool and a few drops of baby oil.

REGULAR INSPECTION

Black spots, dark wax, or redness indicate the presence of ear mites. If you see a liquid discharge, consult your vet. Also look for scratches and bald patches. If your cat has white ears, make sure there are no scabs and sores around the tips of its ears, since this could be a sign of skin cancer.

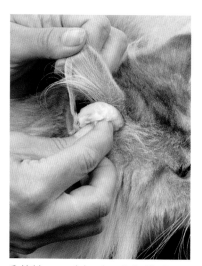

1 Hold your cat's head gently but firmly, and fold back its ear. Using a piece of cotton wool moistened with a few drops of baby oil, wipe away any dirt on the inside of the ear.

2 Cats have very sensitive ears, so be gentle. Do not rub, but instead try to lift any waxy build-up on to the cotton wool. Never use a cotton wool bud, which could cause serious damage to the ear canal.

60 TRIMMING CATS' CLAWS

Outdoor cats usually wear down their claws by walking on paved areas and scratching trees and wooden posts. However, indoor cats need to have their claws trimmed regularly, to prevent them growing so long that they inhibit normal movement or pierce the paw pad. Claws are important parts of your cat's anatomy. Never consider declawing your cat just to solve a scratching problem. Declawing involves the removal of the last joint of a cat's toe. It is a cruel and painful procedure.

Apply pressure on paw pad

Exposed claw

How to expose the claw

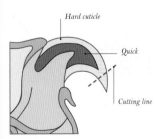

1 Unless your cat is very calm, cut just a couple of claws per session. Holding your cat still, apply a little pressure on its paw pad to reveal the claw to be trimmed.

Hard cuticle

Quick

Cutting line

Cat nail anatomy

2 Using clippers, cut just the white tip of the claw. Be careful not to cut too close to the pink quick. It is better to cut too little than too much, especially if your cat is struggling to get free from your hold.

61 BRUSHING YOUR CAT'S TEETH

The most important thing about brushing a cat's teeth is not to use human toothpaste. Buy a pet-specific product. Before progressing to a toothbrush, try a little toothpaste on the cat's lips, then touch a cotton-wool bud to its gums. You will probably need to hold the cat firmly but gently.

THIMBLE BRUSH
Thimble-style toothbrushes are available for pets. These slip over the end of your finger for improved manoeuvrability in cleaning.

62 BATHING YOUR CAT

Most cats do not enjoy being bathed, but sometimes it is necessary, especially with longhaired cats, which have the hardest time grooming themselves. However, don't try to remove any substance that is not water soluble – oil, for example – without first consulting a vet. Before you start, gather your equipment, and consider asking a friend to help hold the cat. Talk soothingly to the cat throughout to keep it as calm as possible.

1 Place your cat in a bowl with about 10cm (4in) of warm water. Using a shower head or a jug, pour warm water over the cat.

2 When the cat's coat is fully soaked, apply a small amount of cat shampoo. Avoid the cat's face, especially its eyes and ears.

3 Massage the shampoo deep into your cat's coat, all the way down its body. Rinse well, with lots of warm water.

4 Wrap the cat in a towel and pat it dry. You may need to use a hair dryer on a quiet low setting for longhaired cats.

63 LITTER TRAINING

Most cats will take to the litter tray with little difficulty. Litter training requires mostly just good timing on your part. Wait until the kitten looks like it is ready to use the litter tray – it will raise its tail as it squats down – and place it in the tray. Instinct will usually do the rest: the kitten will cover its faeces, and the smell will guide it to the tray whenever the need arises.

EARLY LEARNING
Reward a kitten with praise or a treat when it uses the litter tray. This will encourage repeated use.

CLEAN LITTER
Cats are fastidious animals. Scoop the litter tray regularly to keep it clean and odour-free.

Litter-tray liner

64 CAT-FLAP TRAINING

If your cat is going to be an indoor/outdoor cat, you will need to install a cat flap and train your cat how to use it. The flap should be fitted at about 15cm (6in) from the floor for the cat to be able to use it comfortably. Prop it open at first, and encourage your cat to come through with a toy or a food treat. When your cat is confident using the hole, close the flap and repeat the training process, this time, teaching your cat to push the flap open with its paw or head.

FOOD LURE
A bowl full of your cat's favourite food ought to be incentive enough for it to come through the cat flap.

LEARNING THE HOUSE RULES

65 PREVENTING SCRATCHING

Cats have scent glands in their paws, and in nature they use scratching as a way of marking their territory. However, in the home, this natural instinct can turn into undesirable behaviour when cats scratch furniture and curtains. Prevent this from happening by channelling your cat's desire to scratch towards a well-positioned scratch post. Keeping your cat's claws trimmed also helps.

NIP IT IN THE BUD
If your cat is clawing at a table leg or sofa, cover the part of the furniture that your cat wants to scratch with plastic, then place a scratch post next to it.

Scratch post

66 DEFINING NO-GO AREAS

To enjoy a peaceful cohabitation with your cat, you might need to impose limits on where it is allowed. Although some cats like to sleep with their human family, this could end up disrupting your own sleeping pattern. Keep the door to your bedroom closed, and put your cat's bed in another quiet area of the house.

NO ACCESS
Close your bedroom door if you do not wish your cat to enter. The door ought to be closed at all times, not just at night.

Cats love a cosy human bed

67 DEALING WITH STRAYING

The main step to prevent your cat from straying is having it neutered. This will help curb its territorial roaming instincts. In addition, be sure to provide your cat with a safe and inviting home environment that it will want to return to again and again. Make sure it has enough food and plenty of mental and physical stimulation. Finally, keep your cat indoors after dark.

Tail held high *Determined gait*

QUIET TIME
A chaotic household might lead a cat to seek peace and solitude elsewhere. Create a quiet area for your cat within the house.

FOOD-SEEKING MISSION
Some straying behaviour is related to the need for food. Make sure you give your cat enough to eat.

68 REWARDING GOOD BEHAVIOUR

Cats respond better to rewards than discipline. Rather than reprimanding your cat for unwanted behaviour like scratching the sofa, give it a cuddle or a food treat every time it uses the scratch post. Be sure to provide the reward immediately after the cat displays good behaviour, so it will associate the two.

FAVOURITE TREAT
Some cats enjoy food treats. For others, soothing words, a stroke, or a vigorous play session are better rewards.

69 COMBATING AGGRESSION

It is unusual for cats to behave aggressively towards people. When this happens, it is important to understand the reasons behind it. First of all, take your cat to the vet to rule out ill health. Cats become aggressive when they feel threatened, but some may do so when they are bored, so try to create a stimulating environment for your cat.

Ears folded down

TOWARDS PEOPLE
Pain, fear, and frustration are all potential causes of feline aggression. Understanding the reason is the key to providing a solution.

TOWARDS OTHER CATS
If your cat behaves aggressively towards another cat, it could be due to sharing territory or other resources, such as food.

70 DEALING WITH BEGGING

Cats are unlikely to beg for table scraps the way dogs would. However, when this happens, it is important to put a stop to it before the begging turns into jumping on the table at dinner time. Feeding your cat before sitting down for your own dinner will make it less likely to beg for food. If this fails, shut your cat in a different room while you eat.

Scavenging through rubbish

MANIPULATIVE CAT
Giving in to your cat when it is begging opens the door for more undesirable behaviour.

71 LOOKING FOR A MISSING CAT

Always on the lookout for a warm, cosy place to sleep, a cat can sometimes go "missing" in your own home. It is important to know where its hiding places are and ensure they are safe. Laundry baskets, airing cupboards, wardrobes, and drawers are all fair game to cats. Some have a predilection for washing machines and dryers. Keep the door to these appliances closed, and always check for sleeping cats inside before starting them.

MIAOW ON COMMAND
Teaching your cat to miaow upon hearing its name will make it easier to find it – inside the house or out.

SNUG AND COSY
Cats excel at finding the smallest nooks in which to retire for a little snooze. This kitten has hidden in a school bag.

72 SPRAYING

Like scratching, spraying urine is a way for cats to mark territory. Normally, cats feel secure enough in their home not to display this behaviour. However, at times of increased stress – the arrival of a new kitten, for example – a cat (usually male) might spray around the house. Prevent spraying by learning the telltale signs. If your cat still sprays, clean up the affected area with disinfectant to remove the odour.

Sniffing

ABOUT TO SPRAY
Backing up against the chair, this cat is about to spray: its hindlegs are on tip toes, and its tail is raised and quivering.

45

73

IN OR OUT AT NIGHT?

The days of putting the cat out for the night are gone. At night-time, your cat is more likely to be involved in a fight with another cat or with other predators. In addition, despite there being less traffic at night, it is often faster, representing a serious danger for your cat. For its own safety, keep your cat in after dark. The local wildlife will also benefit from your cat being kept indoors at night.

DARKNESS IS NO DETERRENT

With their incredible night vision, cats will still happily venture out after dusk. But many owners prefer to keep their pet in at night.

74

FIRST TIME OUTSIDE

Make sure your cat is confident in the home before letting it out for the first time, and when you do, supervise its first few trips outside. Open the door, step into the garden, and call it gently, maybe waving a feather-stick toy to encourage it to come out. Do not force your cat outside; it will go when it feels ready.

NERVOUS CAT

Cats like being in control of their territory, so being faced with a vast expanse of outdoor space may be daunting at first. They might display fearful behaviour, slinking around very close to the ground.

EXPLORING

Cats are natural-born explorers, and the nooks and crannies in your garden will provide them with hours of fun.

75 THE HUNTING INSTINCT

Cats are fearsome predators, and the hunting instinct is ingrained deep within them. Even domestic cats, which do not have to worry about their next meal, enjoy hunting. For these well-fed cats, hunting is more about the thrill of the chase. This is one theory for why cats play with their prey before killing it.

PATIENT HUNTERS
Hidden in the tall grass, this cat waits patiently for some unsuspecting prey to cross its path.

STALKING
When a cat spots its prey, it approaches slowly, lowering its entire body into a distinctive gait before pouncing.

76 UNWANTED PRESENTS

If your cat is allowed to go out, chances are it will return one day with a little gift for you – a mouse, a bird, or a little frog, often dead, sometimes alive. As disgusting as this behaviour might seem to you, it is important not to reprimand your cat. This gesture is meant out of love, possibly as your cat's humble contribution to the family table. Instead, praise it for its superior hunting skills, then dispose of the prey quickly while your cat is not looking.

PROUD PROVIDER
This cat gives its offering of a dead mouse the way a mother cat presents her kittens with prey. It is a cat's way of showing love and to nurture its human family.

Dead mouse

47

LIVING WITH A CAT

77 READING NEGATIVE BODY LANGUAGE

Since cats cannot tell us how they feel, it is important to learn to read their body language in order to forge successful relationships. The two most expressive parts of a cat's body are its tail and ears, but its eyes and mouth are also very revealing. While not strictly body language, each of a cat's range of vocal emissions is also key to understanding its state of mind.

Flapping tail

Low growl

GROWLING
A feline growl seems to emanate from the pit of the cat's stomach. A growl is a definite warning sign and often a prelude to a fight.

TAIL SWISHING
Do not mistake a cat's swishing tail for friendliness. Unlike dogs, cats wag their tails when they are angry or nervous.

HISSING
A hissing cat, with its ears back and its teeth bared, makes quite a fierce impression. However, a cat hisses when it feels threatened rather than aggressive.

78 READING POSITIVE BODY LANGUAGE

While negative feline body language has been witnessed at some time or other by anybody who shares their life with a cat, much more familiar is a cat's display of positive body language. After all, most family cats are well loved and cared for, face relatively few threats in their daily lives, and are probably relaxed and happy within their home environment. A happy cat can be recognized by a tail held upright, slowly blinking eyes, and ears pointing upwards.

PURRING

Cats purr when they are content and relaxed. However, when they are nervous or distressed, they might also purr as a way to comfort themselves.

KNEADING

Similar to the movements needed to make dough, this padding behaviour is based on the action used to encourage a mother's milk and shows your cat feels happy and loved.

NUZZLING

Cats have scent glands in their lips and chin. By rubbing their faces against a litter mate or their favourite person, they are effectively claiming it or you as their own.

79

TOYS & PLAY

A vital element of feline development, play time teaches kittens about social interaction – what is acceptable behaviour and what is not, for example – as well as the hunting skills they would need in nature. Playing with your cat strengthens the bond between you. Establish a routine that incorporates 15 or 20 minutes of play time each day. It will keep your cat physically and mentally active – and less likely to develop destructive behaviour.

Play glove with hanging toys

Ping-pong ball

CATNIP TOYS
Some cat toys are stuffed with dried catnip, a herb that creates a hormonal reaction in most cats over six months old, giving them a gentle high.

PING PONG
Cats love to chase fast-moving objects, and a lightweight ping-pong ball bouncing around is the perfect toy. Put it away at the end of the playing session to prevent anyone slipping on it.

SWINGING THINGS
A toy hanging on a self-balancing surface will keep your cat thoroughly entertained as it swats at and catches it, re-creating the feeling of chasing live prey.

Cat is focused

Cat bats with its front paws

Tail helps with balance

POUNCING & PLAY FIGHTING
Kittens hone their hunting skills by play fighting with their litter mates. They will take turns ambushing each other and engage in seemingly ferocious wrestles.

Surprised by ambushing kitten

Pouncing unexpectedly

Waste-paper basket

Feather

Kitten is curious and gentle

HIDE & SEEK
Kittens love playing hide and seek. If possible, provide them with a couple of cardboard boxes from which they can keep an eye on things and ambush at will.

FEATHERS
A feather offers a taste of the excitement of chasing birds. To avoid scratches, use a feather stick, but never leave any cat alone with feathers or with toys that include lengths of string or ribbon.

FELINE FISHING
A dangling toy on a ribbon is a great bait to reel in your cat, which will be unable to resist the quick movement as you flick it above its head. The cat will use all sorts of acrobatic skills to catch its "prey".

80 LIVING WITH A TIMID CAT

Some cats enjoy meeting people, while others run and hide when faced with a new person in the house. This could be because they were not socialized properly as kittens. If your cat is timid, give it constant reassurance by speaking to it in a soft and encouraging voice. Do not force it to face whatever is making it nervous.

SAFE CORNER
Position an igloo or a cardboard box with a warm blanket in it in a quiet corner of the house, so your cat can take refuge in it when it feels afraid.

CURIOUS BUT AFRAID
A cat in new surroundings will investigate cautiously at first. Left to sniff around at their own pace, though, most quickly grow in confidence.

81 MOVING HOME

When you move to a new home, even if it is just to the other side of town, keep your cat inside for the first couple of weeks. The homing instinct is strong in cats, and they sometimes try to find their way back to their old territory. Wait until your cat has accepted the new home as its territory before letting it out.

Wicker cat carrier

SAFE TRAVELS
Home moves can be chaotic. Keep your cat in its carrier, and do not let it travel with the removal people; instead, take it with you.

82 TRAVELLING WITH A CAT

Most cats do not enjoy car journeys, not least because for them it means being confined to their carrier. Do not be tempted to allow your cat, no matter how docile, to roam freely in the car. For its own safety and yours, always use its carrier. Line it with a blanket for comfort, and put a few toys in there. If you are embarking on a long journey, stop regularly to allow your cat to take litter breaks and to have some food and water.

1 Put the carrier in an enclosed room with no escape routes. This is because your cat might put up a struggle.

2 When your cat is in, lock the door, making sure not to trap its paws or tail. A favourite towel or blanket will help comfort your cat.

3 Put the carrier on the back seat, ideally with another person holding it steady. If this is not possible, secure it with a seat belt.

83 CATTERIES & SITTERS

Make sure your cat is in good hands if you are going away. Neighbours can help, but if you are going away for several days, it might be better to hire a cat sitter or to use a cattery. A cat sitter comes into your home daily to feed and play with the cat and to clean its litter. A cattery is like a hotel for cats, with individual pens offering all the comforts of home. A reputable cattery will expect your cat to be vaccinated before taking it in.

THE SAFETY OF HOME
The advantage of hiring a cat sitter is that your cat is able to remain in its own familiar surroundings.

MULTIPLE-CAT PEN
Most catteries have pens for single cats, plus a handful of larger pens for two or three cats that are used to living together.

YOUR CAT'S HEALTH

84 SIGNS OF GOOD HEALTH

If you know your cat and its normal behaviour well, you will be better placed to spot problems early, especially if you also have a regular examination routine. If anything concerns you in terms of the cat's health, call your vet for advice. You may simply be asked to monitor the cat for further developments and to report back. You will probably also be told to take the cat in to see the vet immediately if the condition worsens.

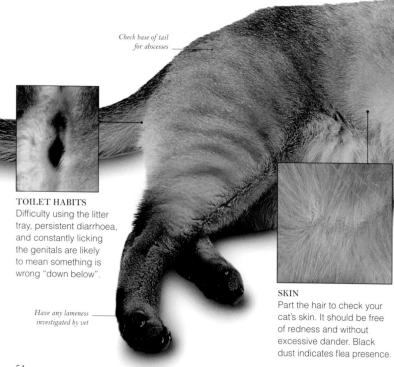

Check base of tail for abscesses

TOILET HABITS
Difficulty using the litter tray, persistent diarrhoea, and constantly licking the genitals are likely to mean something is wrong "down below".

Have any lameness investigated by vet

SKIN
Part the hair to check your cat's skin. It should be free of redness and without excessive dander. Black dust indicates flea presence.

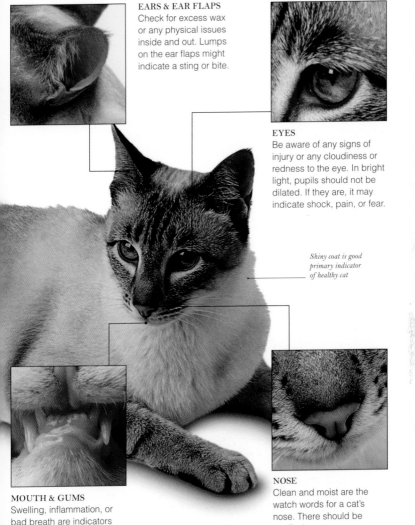

EARS & EAR FLAPS
Check for excess wax or any physical issues inside and out. Lumps on the ear flaps might indicate a sting or bite.

EYES
Be aware of any signs of injury or any cloudiness or redness to the eye. In bright light, pupils should not be dilated. If they are, it may indicate shock, pain, or fear.

Shiny coat is good primary indicator of healthy cat

MOUTH & GUMS
Swelling, inflammation, or bad breath are indicators of underlying problems.

NOSE
Clean and moist are the watch words for a cat's nose. There should be no discharge.

85

CARING FOR AN ELDERLY CAT

A cat that is well looked after can live 15 years or longer. You may need to alter your cat's routine as it enters its twilight years. Its appetite might change, and you may need to feed it more frequently. If this is the case, make its meals smaller. Keep an eye on your cat's toilet habits, as well as on the condition of its coat. Being aware of any changes is the key to catching any age-related problems well in advance.

SLEEPING
Older cats tend to sleep more. They may also be slightly confused upon waking, so try not to disturb them unless necessary.

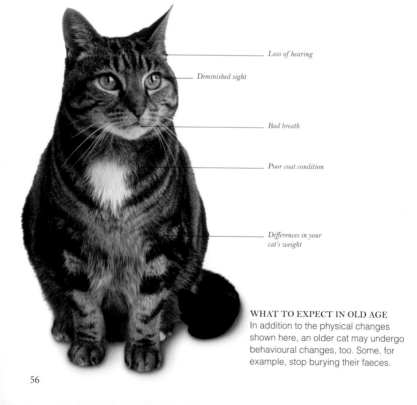

Loss of hearing

Diminished sight

Bad breath

Poor coat condition

Differences in your cat's weight

WHAT TO EXPECT IN OLD AGE
In addition to the physical changes shown here, an older cat may undergo behavioural changes, too. Some, for example, stop burying their faeces.

CARING FOR A PREGNANT CAT

86

Although it is advisable to spay your female cat before her first season, there might come a time when you have to deal with a pregnant cat. A feline pregnancy becomes visible at about five weeks, which is roughly halfway through the term. If you are a responsible owner, your cat should already be vaccinated; if she is not, do not vaccinate her during the pregnancy, since this could affect the health of the foetuses. As the pregnancy progresses, your cat will become less active.

CARE CHECKLIST

A pregnant cat will have an increased appetite. Make sure she is well fed, both during and after the pregnancy. Producing milk is hungry business!

Notice where your cat has chosen to nest, and make this area as comfortable as possible. As the pregnancy progresses, she will spend more and more time there, impregnating the bedding with her smell as a beacon for her soon-to-be-born kittens.

Even an inexperienced cat will probably be fine without your help during the delivery. Just sit back and enjoy the show.

Relaxed face

Round belly

KITTENING BOX
You will often find a heavily pregnant cat, also known as a queen, lying down. This is to distribute evenly the weight in her belly.

PROUD MUM-TO-BE
This queen proudly displays her pregnant belly. Cats usually give birth to between two and five kittens.

87

CONTROLLING FLEAS, TICKS & WORMS

Treat your cat for fleas regularly, especially during the warm summer months. Since fleas can survive away from the cat, it is also important to treat the home environment. Ticks are blood-sucking parasites that latch on to your cat's skin, especially around the face. Worms are internal parasites that can live in the lungs or the intestines; the best preventive measure is a worming tablet every six months.

FLEAS & TICKS

Symptoms: If your cat scratches obsessively, it might have fleas. Part its coat so you can see its skin. Fleas look like tiny brown specks, but the presence of small black droppings is also a sign of an infestation, as is irritated skin. Ticks look and feel like small warts on your cat's skin.

Treatments: Flea treatment is available in the form of several products that are applied to the back of your cat's neck; your home should also be treated. Ticks must be removed manually; it is a tricky task, so ask your vet to explain how.

Flea treatment

EAR MITES

Symptoms: Persistent scratching of and pawing at the ears and vigorous shaking of the head are usually symptoms of an ear-mite infestation. Check the inside of your cat's ears for the presence of dark spots and brown waxy discharge. There might also be an unpleasant smell.

Treatments: Flea-treatment products usually eliminate ear mites, too. In the case of severe infestations, your vet might prescribe drops.

Itchy ears

WORMS

Symptoms: There are different symptoms for different types of worm. The lungworm causes breathing problems, such as a persistent cough, while worms that take up residence in your cat's intestines might cause weight loss, diarrhoea, and anaemia, which reveals itself as pale gums. Worms might also be visible in your cat's stools.

Treatments: Give your cat a worming tablet every six months. Some flea-treatment products are also effective in the treatment of worms.

Vet diagnosis

88 DEALING WITH FURBALLS

Cats are fastidiously clean, and daily grooming means that they ingest a surprising amount of their own hair. Most cats will eat a little grass to bring up a furball. Specially formulated furball-formula cat food is also available. And of course, you can do your bit by brushing your cat regularly.

Self-grooming

ACCUMULATED FUR
Longhaired cats are particularly prone to furballs, because they inevitably ingest a larger quantity of hair while grooming.

89 SPOTTING SIGNS OF DISTRESS

When they are poorly, cats have a tendency to hide and isolate themselves rather than seek help or even comfort from their human family. Although, in many cases, cats are great self-healers, some conditions require veterinary assistance. It is therefore important to recognize the signs that all might not be well.

HIDING AWAY
A normally sociable cat that suddenly takes itself away and stops interacting with its human family might be unwell.

EATING & DRINKING HABITS
Investigate any changes in your cat's appetite and thirst, since the underlying reasons could be serious, such as diabetes.

UNJUSTIFIED AGGRESSION
If your cat behaves aggressively for seemingly no good reason, it could be because it is in pain. Take it to the vet.

90 WOUNDS

If your cat comes back from its outdoor patrols with an open wound, it might try to behave as if nothing is wrong. This is a natural response in an attempt not to show any weakness. If the wound is on its paw, your cat will likely tuck it in under its body to prevent you from examining it. While minor scratches and grazes may be treated at home with a cold compress and a bandage, any open wounds should be seen by a vet immediately.

BANDAGING WOUNDS
It is quite difficult to put a bandage on an injured, stressed cat, so you might want to leave this task to the vet.

91 STINGS

Curious explorers, cats sometimes come up against bees and wasps. If your cat has been stung by an insect, the affected area will appear red and swollen. If you see something that resembles a splinter wound, it could be a bee sting. Try to remove it with tweezers. Some cats are allergic to bee stings, and they might develop respiratory difficulties. A visit to the vet is recommended.

A wasp can give a nasty sting

TREATING BITES
Bites usually need just a course of antibiotics to heal. However, sometimes a bite turns into an abscess. This will require lancing.

92 BITES

Even a neutered cat with a diminished territorial drive will have the occasional dispute with a rival. While scratches tend to heal without medical attention, bites can become infected, because cats' saliva harbours bacteria that may enter the bloodstream. Venomous spiders or snakes may also be a concern in some parts of the world. If you see a puncture wound on your cat's skin, see your vet immediately.

93

WHEN TO GO TO THE VET

Learning to read the signs of ill health in your cat will prevent unnecessary trips to the vet, which can be stressful for your cat and expensive for you. However, do not hesitate to consult your vet if your cat has a persistent complaint, visible injuries, or if it appears to be in shock or in pain.

CHOOSING A VET

The best way to find a vet is by word of mouth. Ask friends and neighbours where they take their pets and whether they are happy with the care they receive there. You can also ask your local rescue centre where they take their residents for check-ups.

At the surgery, ask about their facilities and whether they provide round-the-clock emergency care. Also find out how many vets practise there and whether your cat will see the same one each time.

INSURANCE

Veterinary emergencies are expensive affairs. In addition to the price of treatment, you might have to consider the cost of boarding, since an injured cat may need to be kept in overnight for observation.

To help deal with such bills, insure your cat. You may still have to pay your vet up front, but after filing a claim, you should get a large percentage of your money back.

CONFIDENT HANDLING

Vets have a strong, confident way of dealing with animals, and they know how to pacify cats that are in pain or injured.

EMERGENCY CARE

Medical emergencies can happen at any time. Make sure you know the out-of-hours contact details for your chosen veterinary practice, as well as the location, which may be different.

94

SYMPTOMS OF ILL HEALTH

Although cats cannot tell us what it wrong with them, their bodies and their behaviour speak volumes about their state of health. Anybody who shares their life with a cat will soon get to know what constitutes normal behaviour for their feline companion and what is unusual. Keep an eye out for any changes in their appetite, energy levels, and litter-tray habits. At least once a week, examine their mouth, ears, and eyes, and check their breath. If anything seems out of the ordinary, consider a trip to the vet.

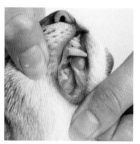

PALE OR INFLAMED GUMS
The colour of your cat's gums is indicative of potential problems. Healthy gums are pale pink.

CHANGE IN EATING HABITS
Monitor your cat's appetite, since any changes could be signs of ill health. If your cat refuses to eat for more than 24 hours, consult the vet.

STRANGE BEHAVIOUR
Cats are creatures of habit. If they suddenly display unusual behaviour – howling when they have always been docile, for example – there might be something wrong.

Be aware of difficulty in litter tray

LITTER PROBLEMS
Keep on top of any bowel issues and urinary-tract problems by cleaning your cat's litter tray regularly. If your cat cries loudly while in its tray, it might have a urinary tract infection.

LETHARGY
A lack of interest in everything is a definite cause for concern in cats, especially if it is accompanied by loss of appetite.

LIST OF OTHER SYMPTOMS
See your vet if your cat displays any of the symptoms listed below or, indeed, any signs of ill health.

• **Increased thirst, frequent urination, weight loss, bad breath, and mouth ulcers** are indicators of chronic kidney disease, common in older cats.

• **Vomiting and dehydration** are more in keeping with acute kidney disease, which is more readily seen in younger cats and often caused by an infection.

• **Diarrhoea** can be caused by a range of problems, as benign as a food allergy or as serious as feline infectious enteritis.

• **Rapid, laboured, or noisy breathing** should be investigated as soon as possible.

• **Lameness** can indicate a large range of potential problems, including hard- and soft-tissue injuries, as well as age-related and more serious concerns.

Scratching with hind paw

SCRATCHING
Persistent scratching is probably related to the presence of fleas, ear mites, and other parasites. Check your cat's coat and ears for uninvited guests.

Drooling saliva

Low posture

EXCESSIVE DRIBBLING
Any unusual drooling is cause for concern. There might be something stuck in your cat's mouth, or it might have ingested a toxic substance. Take it to the vet at once.

95 COMMON CAT ILLNESSES

Just as in humans, there are myriad ailments, illnesses, and complaints that can affect a cat's various body parts. Some are caught from other cats, while others might be common to certain breeds. The key to the best treatment is to spot issues early. This can best be facilitated through routine handling and checking of your cat, and by knowing what feels and looks normal for it.

Longhairs (or Persians) often have breathing problems

JOINTS

Joint disease can be so mild as to go unnoticed by a pet owner or, at the other extreme, can greatly affect an animal's life. Most cases fall somewhere between the two.

PROBLEM SIGNS
• Limping • Stiffness • Inability to jump • Lack of desire to move • Difficulty climbing stairs • Favouring one leg over another • Displaying signs of discomfort or pain when moving

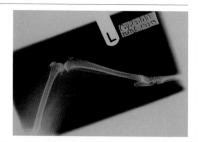

EYES

Among the most common problems with cats' eyes are conjunctivitis, cataracts, glaucoma, keratitis, and a visible third eyelid. Examine your cat's eyes regularly to ensure all is well.

PROBLEM SIGNS
• Redness • Swelling • Presence of discharge • Cloudiness • Wateriness • Opacity of the lens • Bulging eyes • Tear-stained fur around eyes • Inflamed cornea • Visible third eyelid

EARS

Infections of the ear can be uncomfortable and should be treated quickly. Ear mites are commonly found and contagious, so treat all your cats at the first sign of infestation.

PROBLEM SIGNS
• Tilted head • Head shaking • Redness/swelling • Unpleasant odour • Pawing/scratching ear area • Loss of balance • Sensitive near ears • Discharge • Dark brown wax • Hearing loss • Bleeding

SKIN & HAIR

A wide range of problems can affect a cat's skin and hair, including bacterial infections, yeast infections, fleas and lice, ringworm, allergic dermatitis, and even alopecia.

PROBLEM SIGNS

• Constant scratching or licking • Hair loss • Blotchy or red skin • Spots around chin • Red ear flap • Lesions • Chewing obsessively at reachable extremities • Black dust in fur • Flaky skin

DIGESTIVE SYSTEM

There are many ailments that can strike at the digestive system. Some of the most common are intestinal obstruction, food poisoning, and anal-sac irritation.

PROBLEM SIGNS

• Diarrhoea • Vomiting • Frequent regurgitation • Constipation • Distended belly • Loss of appetite • Obsessive cleaning of anal area • Weight loss • Increased frequency of defecation

RESPIRATORY SYSTEM

Cat flu is a chief cause of sneezing in cats, but see your vet if any discharge is produced. Breathing problems can be caused by physical obstructions or injury, among other things.

PROBLEM SIGNS

• Wheezing • Coughing • Sneezing • Choking • Gagging • Rapid, shallow breathing • Noisy breathing • Difficulty breathing

HEART & BLOOD

From heart disease and blood clots, to thyroid problems and anaemia (which may indicate underlying concerns), there is no shortage of issues that can affect a cat's circulatory system. See a vet if you are concerned.

PROBLEM SIGNS

• Lethargy and weakness • Weight loss • Diminished appetite • Fever • Depression • Increased heart rate

GIVING TABLETS & MEDICINE

96

Administering medicine to cats is notoriously difficult because they are usually uncooperative. It is important to get it right first time, since they will become increasingly stressed, which is likely to cause them to lash out with each repeated attempt. If necessary, have somebody help hold your cat still. Mixing crumbled-up tablets with the cat's food is a good idea in theory, but cats have a remarkable sense of smell and will no doubt detect any anomalies in their bowl.

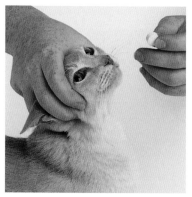

1 Hold your cat firmly, but not roughly, by the head, without touching its whiskers. Place your index finger and thumb either side of its mouth, and tip its head back slightly.

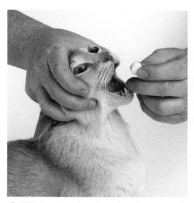

2 With the other hand, gently prise your cat's mouth open by pressing on its jaw, and put the tablet on the back of your cat's tongue. Close its mouth, and massage its throat.

LIQUID MEDICINE
If your vet prescribes liquid medicine, administer it with a syringe. Hold the cat's head still with one hand but do not tilt it back, then slowly squirt the liquid into the side of its mouth.

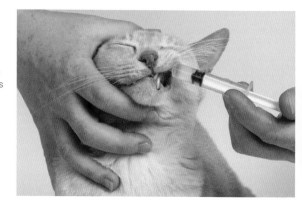

97 GIVING EYE DROPS

Your cat's eyes should be clear of any discharge before you apply eye drops, so clean the area around them with a ball of moistened cotton wool. Squeeze the eye drops into the eye, then close the eyelid for a few seconds, to allow the drops to spread evenly on the cornea. Apply eye drops or eye ointment to your cat's eyes only if you have been instructed to do so by your vet.

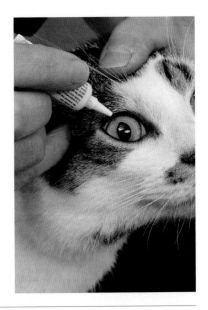

HOLD STILL
Hold your cat's head still in the same way as you would when administering a tablet (see Tip 96). Your cat might struggle, so be sure not to scratch its eye with the nozzle.

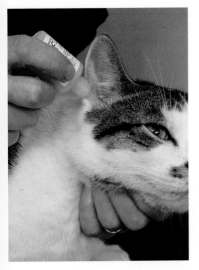

98 GIVING EAR DROPS

If your cat is suffering from an ear-mite infection, it will be prescribed a course of ear drops. Before starting, wipe away any visible dirt from the affected area with a piece of moistened cotton wool. Gently hold your cat's head (see Tip 96), tilting it so that the ear to be treated is pointing upwards slightly. Apply the drops, then massage them gently into the ear.

CAREFULLY DOES IT
As always with cats, it is important to be firm but gentle. Do not insert the dropper into the ears, which are extremely delicate.

99 CARING FOR A SICK CAT

If your cat is poorly or recovering from an injury, it will need a little extra care. Help your cat regain its strength by making a sick bed for it. A cardboard box lined with soft blankets ought to do the trick. Position it in a quiet corner, away from draughts and foot traffic.

Lukewarm hot-water bottle

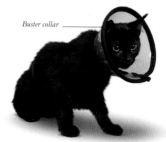

Buster collar

EXTRA COSY
Make your cat's bed more appealing than ever by putting a hot-water bottle full of lukewarm water under a towel or blanket.

SCRATCH PREVENTION
If your cat has received an injury and has been given stitches, a buster collar will prevent it from bothering them.

100 FEEDING A SICK CAT

A sick cat might have to be on a special diet, which it will probably find unappetizing, or it might go off its food altogether. There are a few tricks to get your cat to eat – from serving it its favourite food, to spoon-feeding it nourishing liquid food.

FOOD FOR SICK CATS
A sick or recovering cat needs to be kept well nourished. Try feeding it smaller meals than usual. If your cat enjoys fresh food, such as fish or chicken, prepare some for it. Serve the food warm, since this will activate the smell, making it more pungent and appetizing. If all else fails, you might have to syringe-feed your cat some liquidized high-protein food.

SPOON-FEEDING YOUR CAT
Hold your cat's head in the way shown in Tip 96. Spoon-feed it some unsalted chicken stock or liquidized convalescent cat food.

101 HUMAN HEALTH ISSUES

The most commonplace cat-related human health issue is an allergy. A cat allergy manifests itself with red, itchy eyes, sneezing, and a tickly throat. Some people also develop skin rashes, especially if they get scratched. It is, however, possible to build up a resistance to a cat allergy, and there are also hypo-allergenic cat breeds. Other, more serious conditions include rabies and toxoplasmosis.

ALLERGIES
People who are allergic to cats display a range of symptoms that usually affect the respiratory system.

Beware of cat bites

RABIES
Rabies is carried in some animals' saliva. In countries where rabies is present, even the most innocuous cat bite ought to be taken seriously.

TOXOPLASMOSIS
Cat droppings may contain a parasite that causes a range of diseases, including skin infections and toxoplasmosis. The latter is a greater risk for pregnant women. Wear rubber gloves when cleaning your cat's litter, and wash your hands thoroughly afterwards.

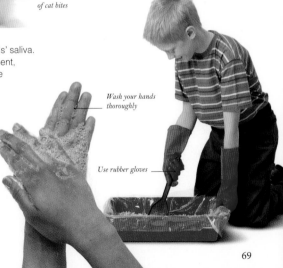

Wash your hands thoroughly

Use rubber gloves

INDEX

ACKNOWLEDGMENTS

Sands Publishing Solutions would like to thank
Kim Bryan for her efficient consultancy work during the project,
as well as for her assistance in fine-tuning the contents at the outset;
Natalie Godwin for design assistance; and the ever-brilliant Hilary Bird
for making such swift work of the index.

Dorling Kindersley would like to thank the following photographers:
Paul Bricknell, Jane Burton, Steve Gorton, Anna Hall, Mark Henrie,
Dave King, Alison Logan, Steve Lyne, Tracy Morgan, Gary Ombler, Tim Ridley,
Paul Self, and Tracy Morgan Animal Photography.

Picture credits
11 cr: Sylvia Tombesi-Walton.
15 r: Cheuk-king Lo. Pearson Education Asia Ltd.
24 bl: kitten courtesy of The Mayhew Animal Home and Humane Education Centre.
42 br: Sylvia Tombesi-Walton.
49 bl: kitten courtesy of Betty.

All other images © Dorling Kindersley.
For further information, see www.dkimages.com